INFORMATION CIRCULAR 9484

GETTING TO ZERO: THE HUMAN SIDE OF MINING

By Elaine Cullen, Thomas Camm, Mike Jenkins, and Launa Mallett

U.S. DEPARTMENT OF HEALTH AND HUMAN SERVICES
Centers for Disease Control and Prevention
National Institute for Occupational Safety and Health
Spokane Research Laboratory, Spokane, WA

March 2006

ORDERING INFORMATION

Copies of National Institute for Occupational Safety and Health (NIOSH) documents
and information about occupational safety and health
are available from

NIOSH-Publications Dissemination
4676 Columbia Parkway
Cincinnati, OH 45226-1998

FAX:	513-533-8573
Telephone:	1-800-35-NIOSH
e-mail	pubstaft@cdc.gov
Web site:	www.cdc.gov/niosh

DHHS (NIOSH) Publication No. 2006-112

TABLE OF CONTENTS

INTRODUCTION[1]

C.M.K. Boldt[2]

The material in this Information Circular was presented at the National Institute for Occupational Safety and Health's (NIOSH) open-industry briefing held during the 2004 Northwest Mining Association conference in Spokane, WA. The open-industry briefing discussed results of recently completed and on-going mine safety- and health-related research conducted at NIOSH's Spokane and Pittsburgh Research Laboratories on the human side of mining—the miner. The first paper sets the stage for what mining has achieved and what we mean by "getting to zero." The second paper, "Help Experienced Miners Become Great On-the-Job Trainers," distinguishes between experienced miners and trainers. This paper provides some key concepts for turning great miners into great trainers. "The Power of Story-Telling in Getting Through to Miners" explains the history and importance of stories in the culture and training of miners. Finally, "Understanding Self in Stressful Working Environments," describes the personal and social aspects of the working environment, much of which parallels familial theories.

The downward trend of deaths in U.S. mining has been remarkable. The number of miners killed or injured each year has decreased steadily over the past 100 years. Some charts show these trends overlain with important events such as World War II and the Mine Safety and Health Act. However, this downward trend in mining fatalities has flattened out. How can you get to zero accidents? Do you concentrate on technological advances and equipment maintenance to avoid malfunctions? Do you concentrate on developing work processes and organizational structures that minimize risk? Do you combine them and take a systems approach that links one to the other with an evaluative loop? Safety efforts in mining have used all of these approaches. What more can we do to get to zero?

Getting to zero will require two objectives: A clear and unshakeable belief that it can be done and attention to the human side of mining. Getting to zero is not a new idea; it has been used by mine companies and safety professionals for years as the safety goal. But getting to zero is not so much a physical goal as it is a belief that there are no insurmountable barriers to achieving the goal.

---

[1]The findings and conclusions in this paper are those of the authors and do not necessarily represent the views of the National Institute for Occupational Safety and Health (NIOSH).

[2]Acting branch chief, Spokane Research Laboratory, NIOSH, Spokane, WA.

ACHIEVING ZERO FOR MINERS' HEALTH AND SAFETY

F. M. Jenkins[3]

INTRODUCTION

Welcome to the 2004 NIOSH open-industry briefing being held in conjunction with the Northwest Mining Association's annual meeting in Spokane, WA. The theme for this year's briefing is a little different. We usually present the results of specific research directed to addressing a particular health or safety problem. This year, we are focusing on NIOSH research that may hold the key for future advances in health and safety for the mining industry. To set the stage for today's session, I would like to present the concept of "zero," which is something being discussed as the next safety goal for mining health and safety.

I first want to make it clear that the idea of achieving zero injuries and occupational illnesses is not a new one, nor is it a new NIOSH initiative. What is noteworthy, however, is that the idea of adopting zero as a health and safety goal for the whole U.S mining industry is gathering momentum, not only in terms of accepting the concept itself, but with the belief that zero is an achievable goal.

What does this vision of zero mean? It means doing whatever it takes to eliminate circumstances in the workplace that harm miners. When you state it that way, you realize that you have to understand and manage a lot of things in the workplace to achieve the zero goal, everything from controlling the working environment to developing a mind-set in miners toward safety.

Mining has always been considered a high-risk industry. As such, injuries and even deaths have historically been seen as an inevitable consequence of mining. This has been the case for many high-risk jobs, including commercial fishing, logging, and steelworking. Miners consider themselves capable of enduring the hazards of mining. They are generally more likely than workers in other industries to accept personal risk or they would not be miners in the first place. To achieve zero, we need to first understand what motivates miners and then figure out how to train and encourage them to work safely.

WHERE ARE WE NOW?

What is the state of the mining industry today? In recent months, mining has taken an economic up-turn, and the future looks brighter. The demand for almost all mining products is high. Fatalities are down, and the statistics indicate that mining is safer than ever.

But if you have worked in the mining industry for very long, you probably have a first-hand experience that reminds you of the risks and possible consequences of mining.

When I was a student, I worked summers in the coal mines for Consolidation Coal Co. My second and third summers, I worked at the Itmann No. 3 Mine doing time studies on one of the first mechanized longwalls in the country. Development was not keeping up with production, and the company wanted to figure out how to speed up the continuous mining cycle used to develop the longwall panels. I spent most of my time with a roof bolter named Teddy McMillion, timing and recording everything he did—tramming, setting up, drilling, and installing bolts. I got to know Teddy and his crew pretty well. After that first summer, I even spent my 2-week Christmas break with them. However, the next year, I decided not to work through Christmas

---

[3]Acting deputy director, Spokane Research Laboratory, NIOSH, Spokane, WA.

break. On Saturday, December 16, 1972, when I was starting that second Christmas break, Teddy and the rest of the crew were riding out of the mine in a trolley-powered portal bus when a spark from the car touched off a methane explosion. Five of the crew, including my friend Teddy, were killed, and three others were severely burned (McAteer, 1998).

For many of us working to make mining safer, these kinds of personal experiences are what make our jobs relevant and the goal of zero very important.

HOW DID WE GET HERE?

In the late 1800's and early 1900's, thousands of fatalities occurred each year in mining (figure 1). Finally society decided that this number was not acceptable. In 1910, Congress created the Bureau of Mines. For 85 years the Bureau conducted research into preventing fires and explosions, reducing roof falls, controlling dusts and gases, and eliminating dozens of other hazards. Mining companies, sometimes with the insistence of organized labor and the government, became more safety conscious. The number of injuries and fatalities dropped.

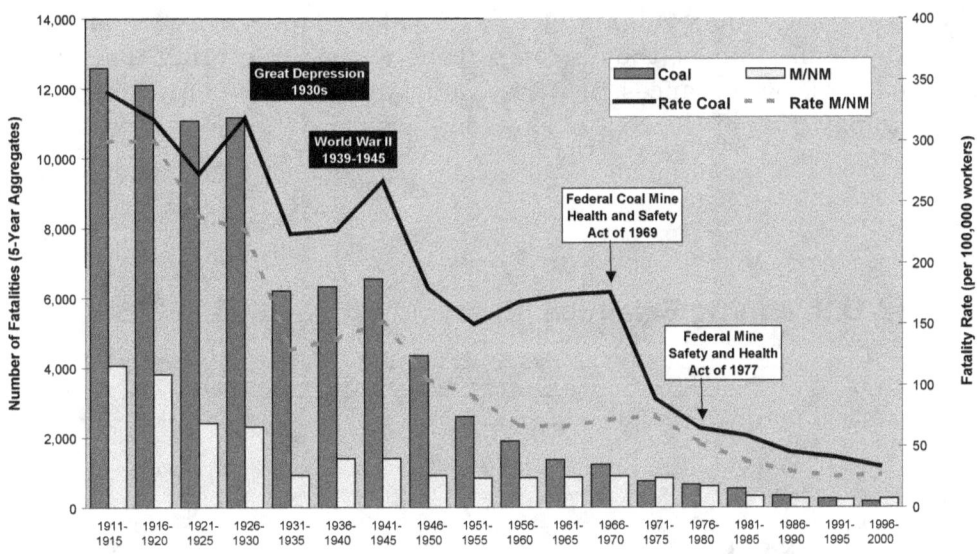

Figure 1.—Number of fatalities (5-year aggregates) and average fatality rates by mining sector (coal versus metal/nonmetal), 1911-2000 (National Institute for Occupational Safety and Health [NIOSH], 2000, 2001, 2004).

By the 1960's and 1970's, annual fatalities had been reduced dramatically, but a series of large-scale disasters led to the demand that more be done to prevent the hundreds of deaths that were occurring each year. Congress enacted the Federal Coal Mine Health and Safety Act of 1969 and later, the Federal Mine Safety and Health Act of 1977.

For the last decade, fewer than 100 fatalities a year have been recorded, but the curve has flattened. More and more safety professionals are asking "What do we do now?" Maybe the answer lies in understanding and addressing why managers and miners are motivated to accept risk.

Risk homeostasis

To explain why it is important to understand what motivates miners, I would like to introduce the theory of risk homeostasis. "Homeostasis" is defined as preserving equilibrium in the presence of ongoing change. Risk homeostasis theory maintains that people accept a certain level of subjectively self-determined risk to

4

their health, safety, and other things they value in exchange for benefits they hope to receive from the activity that exposes them to that risk (Wilde, 1988).

This theory suggests that when people are presented with a less-hazardous way of engaging in the same activity, they will change their behavior in such a way that their personal acceptable level of risk is maintained. The classic example involves antilock brakes (ABS). A 4-year study was conducted in Germany to determine how effective this new braking technology was at reducing the number of accidents in a cohort of cab drivers. In addition, a double-blind experiment was used to determine if the drivers drove differently when their cabs were equipped with the new braking system. The results indicated that cab drivers with ABS changed their driving habits to offset the safety advantages of the new braking systems, that is, they drove faster and shortened braking distances. It was not until they had to pay part of the cost of repairs that accidents were actually reduced (in the final year of the study) (Wilde, 2001).

This example of how humans deal with risk demonstrates that it is not enough to develop safer technology in order to reach "zero"; we also need to understand what motivates miners to take risks and address that as well.

Recent rates

Regardless of the reasons, the trends shown in figure 2 indicate that the rates of deaths and injuries are leveling off in both the East, where underground coal mining dominates, and in the West, where underground hard-rock mines and large open-pit surface operations are more common.

Many leaders in the industry agree that it is time we begin to bring the curve down to a new level and that the only acceptable target is zero injuries and illnesses.

U.S. Mining Fatalities, East and West, 1983 - 2003

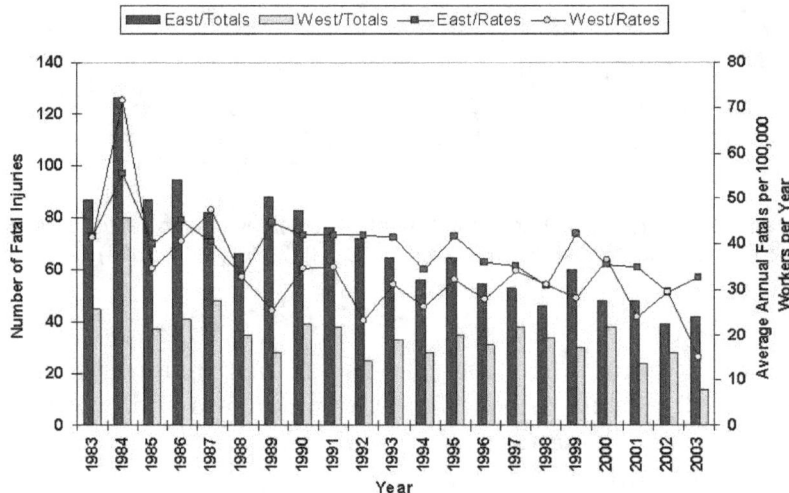

Note: Fatality totals include contractor fatalities. Incidence rates are based on operator fatalities and hours worked only.

Figure 2.—Annual number of fatalities and average fatality rates, east versus west, 1983-2003

ZERO-INJURY CONCEPT

The zero-injury concept is not new. Many examples show how the concept has been applied. DuPont de Nemours is credited with being the original zero injury culture. The company's 11-point safety philosophy begins by saying that "The first and most basic safety principle at DuPont is that all injuries are preventable" (DuPont de Nemours, 1995).

• In the early 1990's, the U.S. Construction Institute established a "Making Zero Accidents a Reality" project team charged with finding out how worker injuries in construction could be eliminated. Its Zero Accident study surveyed 106 of the 400 largest construction companies in the United States and conducted extensive interviews at 38 construction projects. The report identified nine best practices for achieving zero accidents (Pignoet, 2003).

• Mortenson is a large construction company that has implemented a Zero Injuries program. One of its core values, "safety," states that Mortenson is "committed to eliminating all worker injury" (2004).

• Hazard Alert Training, Inc., a group of safety consultants based in Edmonton, AB, has developed a "Zero Injuries Handbook" and provides safety training that incorporates the zero theme (2004).

ZERO AND BEYOND

Figure 3.—Phelps Dodge safety logo

In the mining industry, Phelps Dodge, Inc., has set a very high standard for applying the zero injury concept. The company defines zero to mean "zero incidents, zero injuries, zero fatalities, and zero job-related illnesses." In fact, as the slogan "Zero and Beyond" on its safety logo (figure 3) indicates, Phelps Dodge takes the concept a step further. "Beyond" means reaching the goal and then going beyond it by (1) sustaining zero and (2) taking safety awareness be-yond the workplace into homes and communities (2004).

Last May, I attended a mining health and safety conference in Salt Lake City, UT, co-sponsored by the International Society of Mine Safety Professionals and the Annual Institute for Mine Health, Safety, and Research. I was impressed by the messages of the three keynote speakers, each representing a different part of the industry: enforcement, coal mining, and hard-rock mining. What struck me most was that each speaker had almost the same theme and that a central part of all of their messages included embracing a zero goal.

The first speaker was Dave Lauriski, outgoing Assistant Secretary of Labor for the Mine Safety and Health Administration (MSHA). The theme of his keynote address was that he believes the goal of zero fatalities, injuries, and illness is within reach. He believes that zero should be a common vision for the industry and that, if we work together, we can achieve the goal. He also thinks that we should be more vocal and promote the progress made in mining health and safety. He believes that if we can change the public misconceptions about the safety of today's mining industry, it will empower us and motivate us to move forward to achieving zero. Finally, part of his recipe for reaching zero is to change from a culture of reacting to problems to a culture of preventing problems (2004a).

The next speaker was Brett Harvey, president and CEO of CONSOL Energy and current chairman of the National Mining Association. CONSOL Energy is the largest producer of high-Btu bituminous coal in the United States, operating 19 coal mines in seven states, and is one of the largest U.S. producers of coal bed methane. Mr. Harvey's talk was entitled "Mine Safety: Back to the Future" because he believes that, to achieve the next level of safety, we will have to re-energize proven mine safety methods and techniques while incorporating new concepts that will be necessary to meet the needs of new mining technology and a new generation of miners. Two themes he stressed were accountabil-

6

ity and cooperation. He believes that everyone working in this industry must be accountable for safety and that we all must work together to achieve zero (2004).

The final keynote speaker, Bill Champion, is president and CEO of Kennecott Utah Copper, operator of the largest open-pit copper mine in the world, the Bingham Canyon Mine, which is also the largest man-made excavation on earth. Champion started by saying that "There simply is no better measure of our business success at Kennecott than our safety performance" (2004).

While recognizing zero as the ultimate goal, Mr. Champion sees safety as a journey. He be-lieves that the journey for the mining industry began with efforts to regulate and enforce compliance with newly enacted mine laws. Companies then began to realize that safety was important, both in business and human terms, and began to develop their own safety standards and guidelines. He went on to outline several steps that are part of Kennecott's current journey to zero, things like investing time and energy in training, managing compliance through safety audits, and developing a caring culture. But Mr. Champion closed by saying that the greatest challenge faced by the company's managers and safety professionals is to figure out how to influence employees to make wise decisions, for example, to perform tasks safely.

HOW DO WE ACHIEVE ZERO?

If we accept the charge of these leaders in the mining industry, the next question is, "How?"

As I mentioned before, the U.S. Construction Industry Institute (CII) commissioned a Zero Accident Study (Pignoet 2003). This study identified nine best practices (see table) and identified 170 specific safety techniques being used in the companies they surveyed. These practices were the ones considered "high-impact, zero-injury techniques."

CII High-Impact, Zero-Injury Techniques

1. Demonstrate management commitment
2. Staff ng for safety
3. Safety planning - pre-project/pre-task
4. Safety training and education
5. Worker involvement and participation
6. Recognition and rewards
7. Subcontractor management
8. Accident/incident reporting and investigation
9. Drug and alcohol testing

Of course, these techniques are the types of activities that construction companies themselves can do to promote health and safety.

Likely they include a lot of corollaries for mining companies, and I think it would be a worthwhile effort for the mining industry to conduct a similar study to identify their high-impact health and safety techniques. But I do not believe there are any silver bullets for getting to zero. When we consider the historical advances in mining health and safety, we recognize that mine operators did not make the dramatic improvement on their own.

I believe there is a role for all of us in reaching the next level and achieving zero. I believe we must use a holistic approach. This includes continued efforts by mine operators to value health and safety, by MSHA to update and enforce regulations and provide compliance assistance, by NIOSH and universities to conduct targeted health and safety research, by safety professionals to provide effective training and prevention programs, and by all of us to establish a mining culture that promotes health and safety as a way of life.

TARGETED RESEARCH

At the Spokane Research Laboratory, we are in the process of overhauling our research program. We plan to implement a surveillance cycle model for public health practice. This model includes the timely and accurate gathering of surveillance data with frequent analysis and interpretation; action to control and prevent injuries and exposures through the development and transfer of interventions; and evaluation of our progress using on-going surveillance. In other words, investigate, analyze, provide solutions, and check the results. In this way we plan to make our research as meaningful and effective as possible.

One way of directing our efforts is to find out what is most important to our customers and stakeholders. I encourage you to let us know what you think we should be working on next.

WHAT'S AT STAKE?

Figure 4.—Sunshine Mine Fire Memorial, by Ken Lonn of Auburn, WA, a former miner and shift boss at the Sunshine Mine.

It is important for us to always remember that people's lives are at stake. Figure 4 shows the memorial to the 91 men who died in the Sunshine Mine fire on May 2, 1972, the same year that my friend Teddy died in the methane explosion in West Virginia. As most of you know, this memorial is located only 75 miles from Spokane along Interstate 90 just outside of Kellogg, ID, in the heart of the Coeur d'Alene Mining District. It serves as a reminder that we all have important work to do when it comes to health and safety, particularly those of us in the Northwest.

In his address to this year's MINExpo in Las Vegas, Dave Lauriski said, "It is time to look beyond the 'easy' work of controlling hazards and start looking at the hard work of changing attitudes and behaviors about health and safety. Begin to study and understand the injury and illness patterns in your operation and look at how behaviors can be modified to reduce and eliminate them" (2004b).

In the remaining presentations, we will highlight NIOSH research aimed at understanding human behavior and changing attitudes toward safety.

REFERENCES

Champion, W. (2004). Keeping safety on TRACK: Developing a safety culture. Presentation at International Society of Mine Safety Professionals and the Annual Institute for Mine Health, Safety, and Research (MSHA) Joint Mine Health and Safety Conference, Salt Lake City, UT, May 24, 2004.

DuPont de Nemours and Co. (1995). The DuPont safety philosophy. Available at www.dupont.com/safety.

Harvey, J.B. (2004). Mine safety: Back to the future. Presentation at International Society of Mine Safety Professionals and the Annual Institute for Mine Health, Safety, and Research (MSHA) Joint Mine Health and Safety Conference, Salt Lake City, UT, May 24, 2004.

Hazard Alert Training, Inc. (2004). Retrieved December 5, 2004, from http://www.hatscan.com/ towards-ZERO-injuries-comply.asp.

Lauriski, D. D. (2004a). Institutionalizing a culture of prevention. Presentation at International Society of Mine Safety Professionals and the Annual Institute for Mine Health, Safety, and Research (MSHA) Joint Mine Health and Safety Conference, Salt Lake City, UT, May 24, 2004.

Lauriski, D. D. (2004b). Getting to zero: Stakeholders role in making an industry free of fatalities and injuries. Presentation at MINExpo, Las Vegas, NV. Retrieved December 3, 2004, from http://www.msha.gov/MEDIA/SPEECHES/2004/09282004b.asp.

McAteer, J. D. (1998). Historical summary of mine disasters in the United States, Vol 2: Coal mines–1959-1998. Morgantown, WV: National Mine Health and Safety Academy, MSHA, pp. 77-79.

M. A. Mortenson Co. (2004). Values. Retrieved December 5, 2004, from http://www.mortenson.com/about_us/about_us.html

National Institute for Occupational Safety and Health (NIOSH) (2000). Worker health chartbook. DHHS (NIOSH) Pub. 2000-127, 162 pp.; (2004) Worker health chartbook, DHHS (NIOSH) Pub. 2004-146, 227 pp.

Phelps Dodge Corp. (2004). Zero and beyond. Retrieved December 5, 2004, from http://www.phelps-dodge.com/Health_Safety/ZeroAndBeyond.htm.

Pignoet, V. S. (2003). Achieving zero injuries: Caring is not enough. *Insulation Outlook*. Retrieved December 5, 2004, from http://www.insulation.org/articles/article.cfm?ID=IO031103.

Wilde, G. J. S. (1988). Risk homeostasis theory and traffic accidents: Propositions, deductions and discussion of dissension in recent reactions. *Ergonomics* 31, 441-468.

Wilde, G. J. S. (2001). Target risk 2, pp. 113-115.

HELP EXPERIENCED MINERS BECOME GREAT ON-THE-JOB TRAINERS

Launa Mallett[4]

Preparing new miners is always important, but in some sectors of the mining industry, this issue has become a critical concern. As baby boomers move toward retirement, mine managers are developing strategies to prepare for the impending (and in some cases, current) knowledge and experience gaps. Effective and efficient on-the-job training (OJT) becomes increasingly important in such environments.

A team at the National Institute for Occupational Safety and Health's (NIOSH) Pittsburgh Research Laboratory (PRL) began studying the changing demographics of the mining workforce in 2000. The team found that various segments of the mining industry and mines in different parts of the country had workforces with different age profiles. Overall, however, the mining industry was seeing the same aging being experienced by the rest of the workforce. In 2004, the median age of U.S. workers was 40.5 (Bureau of Labor Statistics, Current Population Survey). The same survey reported that the median age of the U.S. workforce was 45.9 in coal mining, 43.7 in metal ore mining, and 42.8 in nonmetallic mineral mining and quarrying. Clearly, a large number of miners will be leaving the workforce over the next 10 to 15 years.

The research team also looked at what was required to bring new individuals successfully into the mining workforce. It talked with human relations personnel, safety and health specialists, supervisors, experienced miners, and new miners. It found that while OJT was of high interest, little standardization of OJT programs had taken place. Generally, an experienced miner was selected and asked to show the new person how things were done, and no formal selection criteria were established for instructors, few standardized training materials were available, and little evaluation of the training was conducted. An informal system of OJT, which some researchers have dubbed the "Follow Joe" method, was the most common practice. Miners who had experienced this type of OJT relayed stories of great successes and failures.

An informal OJT program is sometimes appropriate. It works best when a small number of people need to be trained. When one or two new individuals are brought in to work with an experienced crew, the new workers are trained not only by the designated OJT trainer, but also by their experienced co-workers. This requires a work crew open to helping the new employee and a designated trainer with appropriate work skills and some teaching skills. An informal method can work for a large operation when it is not important whether or not a consistent message is given to every new person doing a given job. Since trainers don't usually receive instruction related to how or what to teach, they create their own methods and materials. These may vary greatly from trainer to trainer, but if the way the task is completed is not important, then a problem exists only when the end product does not achieve a standard. Informal OJT is also a low-cost way to get information from experienced workers to new employees.

However, informal training is not effective or efficient if (1) many new employees are hired, (2) employees are expected to work with people other than those who originally trained them, (3) a standard method for completing tasks is expected, (4) speed of getting new employees skilled at their jobs is important, (5) a new task or process is to be introduced, (6) experienced workers have developed methods or habits management does not want passed on to new employees, and (7) the teaching experience and expertise of experienced workers are low or unknown. In any of these situations, a formal OJT program is needed.

[4]Sociologist, Pittsburgh Research Laboratory, NIOSH, Pittsburgh, PA.

FORMALIZING ON-THE-JOB TRAINING

Based in the research discussed above, a model for an OJT program was developed. The key to this model is in providing coaching skills instruction to experienced workers who will be teachers. Some experienced miners may be skilled at sharing what they know, but others will not naturally be able to convey the information and skills needed to do a job. Still, in both cases, experienced miners can be taught coaching skills that will improve training.

A coach cannot, however, create a successful program alone. A successful OJT program is created in a team environment. Roles and expectations should be made explicit for the coach, the trainee, the coach's supervisor, the trainee's supervisor, and the program leader or champion. In a trainee-centered approach, more than one coach may be assigned to a trainee. Perhaps an equipment operator and a mechanic who maintains that type of equipment should share responsibility for coaching a new operator. It's up to the trainee's supervisor, with assistance from the program leader, to determine who can best help the trainee learn the new skill.

Another important component to a formal OJT program is standardized methods for jobs or tasks. Coaches must know the standards to which trainees are required to perform. They should know when tasks must be done exactly the same way each time by each individual and when variability is acceptable. The process expressed in the Mine Safety and Health's Job Task Analysis is one way to use the knowledge of experienced personnel to standardize jobs. (See details about the process and contacts at http://www.msha.gov/interactivetraining/task training/index.html.) Regardless of the process used to standardize jobs, the methods for each job must be reviewed and, when needed, revised to continue to be helpful. Training materials should be based on these job standards. It is best if all coaches for a given job teach from the same materials so that the same message is being learned by everyone being taught a specific job or task.

It is management's responsibility to create an atmosphere in which a formal OJT program can succeed. In such an environment, coaches will be recognized not only for their task knowledge and skill, but also for their coaching knowledge and skill. They will be given training to develop and continue enhancing their training abilities. Supervisors and other employees will believe that time given to training workers new to a given task is time well spent. They will understand that a short-term loss in production during training will be compensated for by the effectiveness of the well-trained employee over a longer term. It is especially important for management to support training for new employees as they develop their understanding of the organization's culture and their places in it. If new employees are trained well, they will expect to provide good training to others when it is their turn to be supervisors and coaches.

Pairing a new miner with an experienced miner is a time-honored way to pass on skills and knowledge.

SELECTING A COACH

Because coaches are the key to an effective OJT program, their selection is one of the most important parts of program development. So what is a coach? A coach has been said to be "someone who helps someone else [a trainee] learn something that he or she would have learned less well, more slowly or not at all if left alone" (Chip Bell). Coaches don't do magic by giving knowledge or skill that could not have come from any other source. Instead, they improve the efficiency of the system by shortening the learning period for a worker new to a given task. Some people have natural skills at teaching, some have experience from past jobs or from off-the-job activities, and others may not have tried coaching at all or have tried with less-than-positive results. Developing a successful group of OJT coaches starts with selecting the right people for the job.

It may seem that the easiest way to select a good coach for a given job is to find the employee who is best at doing that job. This is not necessarily the case. Technical expertise does not always lead to an ability to share that expertise with others. Sometimes the opposite is true. Experts often see jobs in chunks or patterns rather than perceiving each specific step as they perform it. They may have difficulty breaking a job down into enough simple steps so that a novice can follow and understand. An expert may no longer see the complexity of the work. What is needed is a coach with a good understanding of how to do the job and the ability to share this knowledge.

After the initial criterion of task or job proficiency comes characteristics that set excellent coaches apart from adequate ones. Strive for coaches who exhibit the qualities below.

Desire
Successful coaches want to coach. They take pride in sharing what they know. Improving their teaching skills is important to them. A successful OJT program will have coaches teaching jobs in standardized ways with an emphasis on safety.

Responsiveness
Successful coaches need to engage trainees. Their listening and communication skills are key to conveying the right information. They use questions to direct the learning process. Successful coaches understand that learning will increase when trainees are comfortable asking questions.

Enthusiasm
Successful coaching takes energy. OJT sessions take time to prepare and commitment to conduct well. Coaches who feel good about their jobs will pass on those feelings.

Humor
Successful coaches have a good sense of humor. Things don't always go according to plan. Everyone has learned something the hard way. Laughing about things that have gone wrong sends the important message that messing up while learning is OK.

Sincerity/Honesty
Successful coaches don't fake it. They truly care about the success of their trainees and deal with them in a straightforward manner. Trainees will respect a coach who admits not knowing something and then goes and f nds the answer for both of them.

Flexibility
Successful coaches are adaptable. They know when to eliminate, adjust, or change what they are teaching to match the capabilities of the trainee or outside constraints.

Tolerance
Successful coaches are open to the opinions of others. They recognize and accept differences in personalities. They are interested in what others have to say, even when they don't agree, and accept negative feedback as a tool for improvement.

Commitment
Successful coaches improve over time. Training, practice, and honest reviews help coaches develop and ref ne their skills. Formal certif cation programs provide a public statement of support from the company. Successful coaches take their coaching duties seriously.

12

Every coach will be stronger is some areas than in others. After being selected, each coach can improve by focusing attention on qualities he or she finds the most challenging to exhibit.

A coach can learn how to focus by attending professional development workshops, by reading related books or articles, through discussions with other coaches and trainees, and by practicing their skills.

COACHING WORKSHOP

One means of improving the knowledge and skills of coaches selected to conduct OJT is a 1-day workshop. Materials for such a workshop have been developed by NIOSH researchers and an industry mine safety professional. The workshop was tested with experienced miners who had been selected to be OJT coaches. Following the workshop, miners reported that it was relevant to their jobs and that they would recommend the workshop to others. During train-the-trainer seminars, workshop materials were introduced to mine training and human resources professionals, who were also very positive about the potential for using the workshops to improve their OJT programs. The participants particularly appreciated the highly interactive nature of the workshop.

While these workshops were designed to cover a single day of instruction, they could also be presented in a series of shorter training sessions. They can be delivered as written or modified to meet the requirements of individual workshops and present site-specific examples. A workshop contains the following units:

Welcome and Introductions–The workshop starts with a discussion of why the workshop is being conducted and expectations of both the instructor and the participants. A company manager should explain the importance of the program to the organization.

Unit 1: What Is a Coach?–In unit 1, participants get the opportunity to do a self-assessment to determine their coaching strengths and weaknesses. The characteristics of a good coach are defined.

Unit 2: The Coach/Trainee Relationship–Communication skills are highlighted in this unit. The responsibilities of coaches, trainees, and supervisors are also explored.

Unit 3: Coaching Adults–This unit gives information about how adults learn and strategies for improving teaching them.

Unit 4: Preparing a Training Outline–Using MSHA's Job Task Analysis process, participants prepare training materials and learn how important it is that materials be consistent for each coach who will teach a specific topic.

Unit 5: Coaching Practice–This unit provides an opportunity for practicing the skills learned throughout the day. Evaluation forms for coaches, trainees, and observers (introduced in earlier units) are used so participants have a chance to try their skills and evaluate them under the direction of the workshop leader.

Unit 6: Wrap-Up/Summary–The last unit gives participants a chance to review the day, ask questions, and think about what comes next. The coaches leave the workshop with a plan for using and continually improving their new skills.

After completion of the workshop, the coaches should know where to go for additional resources and support. A follow-up session should be held after they have had the chance to practice their new coaching skills in the workplace.

GETTING STARTED

For ideas about how to develop or improve an OJT program, read *Coaching Skills for On-the-Job Trainers*, NIOSH Information Circular 9479 (NIOSH Pub. No. 2005-146), 2005. References to other documents about training and adult learning can be found there also. Also see "Considerations in Training On-the-Job Trainers," by William Wiehagen, Don Conrad, Tom Friend, and Lynn Rethi in *Strategies for Improving Miners' Training*, NIOSH Information Circular 9463 (NIOSH Pub. No. 2002-156), 2002.

ACKNOWLEDGMENTS

This paper is based on work conducted by Launa Mallett, Ph.D.; Kathleen Kowalski-Trakofler, Ph.D.; Charles Vaught, Ph.D., CMSP; William Wiehagen, CMS; Robert Peters; and Peter Keating.

THE POWER OF STORYTELLING IN GETTING THROUGH TO MINERS

Elaine T. Cullen[5]

They may forget what you said, but they will never forget how you made them feel. (Buechner)

Mining has historically been a dangerous business. While 1907 was a particularly deadly year for the nation's coal miners, with 3,242 fatalities, Mine Safety and Health Administration (MSHA) statistics show that the numbers weren't much better in subsequent years. It wasn't until 1946 that the number of recorded coal-mine deaths fell below 1,000 (www.msha.gov). Unfortunately, this number doesn't reflect the full extent of mine-related injuries and fatalities, since comprehensive statistics weren't kept back then for miners working in the non-coal segments of the industry. (For example, sand and gravel workers were not included until 1958.) The good news is that by 2003, the total number of fatalities among coal and non-coal miners combined was only 56. That number looks pretty good until one considers that every statistic is a worker who did not go home at the end of his or her shift.

Since 1977, when the Mine Safety and Health Act (known in the industry as "the Act") was expanded to include metal/nonmetal mining under its regulations, every underground miner in the United States has been required to attend 40 hours of new miner training before beginning their career (24 hours for new surface miners) and 8 hours annually thereafter. While safety trainers and regulators refer to this as annual refresher training, the miners themselves have been known to call it "safety jail." They traditionally do not look forward to what they consider a wasted day. They are also resistant to having trainers who, in their opinion, are not real miners tell them how to do their jobs, even if the message is to work more safely.

In 1998, after a series of stakeholder meetings held across the West by NIOSH's Spokane Research Laboratory (SRL) indicated that safety trainers perceived a need for better instructional materials, NIOSH funded a pilot project to examine what "effective" training would look like to the mining industry and whether NIOSH could create training that would break through

Waiting for safety training

---

[5]Communications chief, Spokane Research Laboratory, NIOSH, Spokane, WA.

the resistance miners had to formal training. This project was titled "Development and Evaluation of Effective Safety Training for Mining." In 1998, SRL had no prior experience in training development, so the project began with a study of the miners themselves to determine what effective training would look like to them. Personnel from SRL put together a loose-knit group of advisors from the mining industry to create a list of topics that would be appropriate—and needed—for training development and made the decision to limit these materials to the Western noncoal mining industry. The Pittsburgh Research Laboratory in Pittsburgh, PA, was doing a good job in producing training materials for coal miners; therefore, the SRL project would focus on the noncoal segments of the industry.

Mining has a unique culture that helps its members determine how to make sense of their world. Patton (2002) believes that culture provides the road map by which members of that culture negotiate the world.

> Culture is that collection of behavior patterns and beliefs that constitutes standards for deciding what is, standards for deciding how one feels about it, standards for deciding what to do about it, and standards for deciding how to go about doing it. (p. 81)

Van Maanen and Barley (1984) suggest the existence of "occupational communities" (or occupational cultures), which they define as—

> a group of people who consider themselves to be engaged in the same sort of work; whose identity is drawn from the work; who share with one another a set of values, norms and perspectives that apply to but extend beyond work related matters; and whose social relationships meld work and leisure. (p. 287)

Miners share a culture that is based on a long history. Hard-rock mining in the United States has its base in Europe, particularly in those countries with an active mining industry, such as Germany, Sweden, Wales, and Norway. Miners who came to the young country to work in its mines brought their customs, their norms, and even their shared occupational language with them. Even today, the language used in underground hard-rock mines reflects this history and can be incomprehensible to outsiders. Terms such as "grizzly," "stope," "buzzies," "toe," "back," and "nipper" are not commonly understood in the nonmining world. Thus, the language itself is a powerful aspect of an occupational culture that seeks to communicate internally, yet exclude those who do not belong.

Beyond a general mining culture, the Western hard-rock segment of the mining industry has its own idiosyncrasies. Because of a compensation system known as "gypo," or "contract," mining, good miners can make excellent wages, but they are expected to operate in a semi-autonomous fashion. Miller (1991) explains the gypo system and the benefit it brings to mine operators who understand that these expert miners are some of their most valuable assets.

Gypo miner

In addition to being highly productive, however, gypos are notoriously independent, knowing that they can "tramp" whenever they want, leaving their current jobs to move to another mine with higher pay or better benefits, because they know the value a master miner has in this environment and that they can always get a job. Miners in this sector describe themselves as hard working and hard playing. They believe that they earn the right to spend their money as they wish, although they are generally adamant that taking care of their families is a first priority.

In my study of the Sunshine Mine fire of 1972 (Cullen 2004), I describe the culture of hard-rock gypos. These workers are proud of what they can do; they are very good at what they do; and one of the things they truly resent is someone else coming in and telling them how to do their work, especially if that person is not even a real miner. This resentment was an obvious complication to the new training development project, which was unmistakably a government research project. For the project to be successful, it was necessary to downplay that aspect and win the respect and cooperation of the miners.

Because mining has a strong occupational culture, it also has well-established norms for behavior. Sanctions for violating these norms can come in the form of targeted horseplay, of attempts to embarrass the rule breaker, or, for more serious violations, by ostracism. Miners are well aware that certain norms are in effect because they provide some protection against injury and death, and that all safety regulations are "written in blood." Allowing a co-worker to be sloppy or careless puts everyone in jeopardy, so members of the mining culture generally are quick to express their disapproval.

Experienced miner

If occupational culture is a gatekeeper that provides its members with guidelines on what to do and how to do it, then it is also the primary key to successfully changing members' behaviors. However, occupational cultures that are particularly strong because of shared dangers faced by their members (such as police, firefighters, loggers, fishers, or miners) will be very resistant to changes suggested by outsiders. If changes are accepted and promoted by insiders, they are much more likely to be adopted.

> Danger...invites work involvement and a sense of fraternity....Recognition that one's work entails danger heightens the contrast between one's own work and the safer work of others, and encourages comparison of self with those who share one's work situation. Attitudes, behaviors, and self-images for coping physically and psychologically with threat become part of an occupational role appreciated best, it is thought, only by one's fellow workers. (Van Maanen & Barley, 1984, p. 301)

This shared sense of danger and responsibility was a strength that could be used in developing training materials, if the miners themselves could be convinced to participate in creating the new videos.

Many social researchers have studied the distinctive challenges that teaching or training adults present. Knowles, considered the father of adult learning theory, suggests that for learning to take place, adults—

- Need to know why they should learn the information.
- Need their training to be self-directed to some degree.
- Need training to be related to prior experience.
- Must be ready to learn.
- Must be motivated to learn.
- Need to believe that the new information is going to help solve problems they have encountered in the past and will likely encounter again (Knowles, Holton, & Swanson, 1998).

Another leading theorist, Albert Bandura, developed what he calls Social Learning Theory (SLT) and suggests that people learn not only through their own experiences, but also learn by watching others and then copying their behavior if—

- The person they are copying (the "mentor") is similar to themselves.
- The mentor is admired.
- The behavior observed is perceived to have value (1977).

Wlodowski (1985) studied the question of how adults are motivated to learn and concluded that motivation will occur if—

- They believe they will be successful in learning.
- They believe they have a choice in whether or not to learn the information.
- They see the training as valuable.
- They are enjoying the experience.

Finally, in discussing whether or not adults will be successful in learning new information, Zemke (2002) proposes four key elements that must be present.

- Learners must be willing to pay attention to what they are watching or learning.
- Learners must remember what they saw.
- Learners must be physically and intellectually capable of performing the task or using the information presented.
- Learners must be motivated to model the behavior they have observed.

The challenge for the new NIOSH research project, then, was well defined with the help of these social scientists. If the training videos developed were going to be truly effective, we had to find a way to—

- Get the miners to pay attention in their training classes.
- Convince them that the information was important and relevant.
- Make the learning memorable.
- Make the learning enjoyable.
- Motivate the miners to choose to learn.
- Tie all training lessons to real-life situations.

Several of the social researchers who provided a conceptual framework for this project argue for the importance of paying attention to how people feel about their training. No training can be effective if the participants aren't interested, they believe. They agree that it is important for adult trainees to enjoy their training and to find it interesting enough to listen so that they will remember it. A critical element in achieving these objectives is to break through the barriers the trainees have erected.

One means to transfer information in a way that engages interest is through stories. While it is true that stories evoke different memories and responses in different people (generally depending on individual experiences and values), Gargiulo (2002) states flatly that "The hallmark of intelligence is our ability to collect stories and regularly reflect on them in order to gain new insights from them" (p. 6).

The roles played by stories are numerous. Among them, Gargiulo lists the following:

- Stories empower the speaker.
- Stories create an environment of trust.
- Stories create a bond among those who hear them.
- Stories engage the mind.
- Stories have a unique ability to defuse conflict and differences of opinion.

Sharing stories

18

- Stories encode a lot of cultural information.
- Stories provide a way to learn from personal or vicarious experiences.
- Stories can be used as weapons.
- Stories can help facilitate healing.

Miners are natural story tellers. They use stories very effectively to provide information to inexperienced workers. Older miners talk about people they knew, people they worked with or learned from, people who were hurt or killed in the mines because of a moment's inattention or a bad decision. New hires don't need to experience the events in these stories to learn from them, nor do they have to defend their own actions or errors in judgment, because the stories are not about them. Older miners use stories to illustrate what's important to know, how to do things correctly, and what can happen if you don't follow these norms. In doing so, they make the lessons come alive so that they are much more interesting and memorable.

New employees coming into a mine can be overwhelmed with information. This is a complex, potentially dangerous work environment for the uninitiated, and often the new hire is bombarded with unfamiliar rules, terms, and expectations. Stories provide a way for people to reorganize information and make sense of it, as well as making it much easier to remember and determine why it's important.

> "Story" is a way of knowing and remembering information—a shape or pattern into which information can be arranged. It serves a very basic purpose; it restructures experiences for the purpose of 'saving' them. And it is an ancient, perhaps natural order of mind....By imposing the structure of a story onto some circumstance or happening, greater coherence and sensibility are achieved within the event itself, and otherwise isolated and disconnected scraps are bound up into something whole and meaningful. (Livo & Rietz, 1986)

Once we realized that stories were a good way to transfer safety information to miners, the question became "how do we tell a story to more than one person at a time?" The answer, of course, was "through video."

The project funded by NIOSH in mid-1998 has been the vehicle for the creation of eight training videos to date. These videos made use of several of the constructs that were already in use in the mining industry, namely engaging a master miner as a mentor or as the teacher and the use of storytelling as a method of providing information. They also followed accepted principles for teaching adults, in that the videos were related to real situations, shot in real mines, and addressed specific problems. In some cases, stories told by individuals were captured on camera and embedded in the video, whereas in others, a story line was created that allowed us to teach certain lessons as the story unfolded. Simmons (2001) calls these "teaching stories," which "help us make sense of new skills in meaningful ways. You never teach a skill that doesn't have a reason why." The stories in the videos, whether captured or created, all focus on particular hazards or dangers, providing the learner with the opportunity to place him- or herself vicariously in the story to determine how he or she would have behaved in that situation.

The videos that have been produced so far include—

- *Handling Explosives in Underground Mines* (1999)
- *Miner Mike Saves the Day—or—Ground Support...It's Important* (2000)
- *Hazards in Motion* (2001)
- *Hidden Scars* (2001)
- *Zen and the Art of Rock Bolting* (2002)
- *You Are My Sunshine* (2002)
- *Aggregate Training for the Safety Impaired* (2003)
- *The Sky Is Falling!* (2004)

Another video, *Preventing Rock Fall Injuries in Underground Noncoal Mines*, was produced under a different project by Art Miller of SRL and released in 1999.

When the project was initially funded, it was widely believed that miners would not agree to work with us. Because the gypo system required them to work hard and fast in order to earn production bonuses, any time they spent working with us would take time from their work and jeopardize the likelihood they would "make the round." The fear that we would find resistance to the project proved to be unfounded, however. What we had forgotten to consider was that the mining industry had long relied on the master-apprentice relationship to train new miners, and any miner who would agree to be portrayed in the videos as "the master" would have bragging rights among his colleagues. These are very proud people, and "being in the movies" was definitely viewed as a status symbol among them. Rather than begging people to work with us, we encountered many situations where we had more would-be actors than we could use. We did our best to be fair, however, and to move around to different mines so that we could use as many miners as possible.

A research project is, by its nature, about discovery. We learned much over the course of this project about the technical aspects of putting together movies, but also about how to use stories most skillfully. The use of a master miner (in one form or another) as a teacher was fairly constant, but the portrayal of the young miners who represented the learners changed over time. The concept of the "transitional character" was tried in the video *Aggregate Training for the Safety Impaired*. In this video, two new hires, Ted and Slick (who happens to be a crash-test dummy) bungle their way through their jobs and are fired from every mine in the county by the end of the week.

However, these characters learn to be smarter, safer employees. By the end of the video, they are able to articulate what they had done wrong and what they had learned. This construct proved to be quite powerful. Not every miner can hope to become a master, and while most miners recognize a master when they see one, achieving this status may be out of reach for the majority of people.

When we did the initial technical reviews of the aggregate video, however, we were surprised to hear how often reviewers laughed and said that they, too, had done the unsafe actions engaged in by the characters Ted and Slick portrayed. The miners identified more closely with the new hires than with the experts. While they recognized that they were lucky to have survived, in many cases it clearly resonated with them that Ted and Slick got smarter, just as at some point in their careers so had they. Any miner who had survived in the industry for very long, they believed, had started out dumb and dangerous and had become smarter and safer, just as the characters had. The video provided new information in a humorous way for new employees, but for the older, more experienced employees, it was a reminder of what they had been and what they had become.

One of the research questions addressed in this project concerned mentors. The mining industry is aging, and many experts are facing retirement in the near future. The advisory team of safety trainers voiced a concern common in the industry—when these people leave, who will train new miners and keep them alive long enough to become productive and safe workers?

We were asked whether or not it would be possible to capture the wisdom and expertise of these master miners and, using video, train a new generation of miners after the older miners had retired.

Master miner and apprentice

The video *Zen and the Art of Rock Bolting* addressed this question. The video follows an expert gypo, Jim Mortensen, as he teaches a young worker about mining and mining culture. Jim talks about safety, about his long career, and about mining, and clearly communicates (in his own gruff way) why it's important to work safely. "You don't make no money if you're in the hospital," he says.

Evaluations of this video (Fein 2003) showed that it is indeed possible to capture a master miner. When asked the question whether or not the viewer would like Jim to be their teacher, nearly 100% of the people interviewed, regardless of whether they were inexperienced or experienced, said yes. Jim is charismatic, and he is also good at what he does. His personality undoubtedly played a major role in the success of the video, but there are many miners like Jim in the mining industry who could also be captured on video and used to teach future miners.

Several years after the inception of the project, the advisory team of safety trainers asked if we could put together training materials that could be used during mine rescue training. I knew that the Sunshine Mine, a large underground silver mine about 70 miles from SRL near Kellogg, ID, had been the site of one of the worst hard-rock mine disasters of the 20th century. On May 2, 1972, the unthinkable happened, and the mine caught fire, killing 91 men. Although it had been 30 years since the fire when I began this part of the project, the Sunshine disaster was well-known in the industry. Consequently I decided to use the Sunshine Mine as a backdrop for a mine rescue video.

One of the miners starring in the *Hazards in Motion* video, which was in production at the time, was Don Caparelli, a Sunshine shift foreman and a veteran of the Sunshine disaster. Don arranged for me to interview 27 people who had been involved in the fire. Some had escaped from the mine, some had worked on the rescue and recovery crews, and others were community or family members. Both of the only two men found alive inside the mine 8 days after the fire started were also interviewed.

The video that resulted from these interviews, *You Are My Sunshine*, was a much larger project than anyone initially expected. In fact, it became a full-length documentary of the fire and the impact it had on the lives of the people of the Sunshine Mine and the Silver Valley. *You Are My Sunshine* tells the story of the 13 days of the disaster through the combined stories of those who lived it.

A Story of the Sunshine Mine Fire

This is a difficult video to watch. The emotions of those who survived, as they talk about those who did not, are still raw even after 30 years. The Sunshine fire was a turning point in mine safety, however. This fire was the catalyst that finally changed federal law to create "the Act" that brought metal and nonmetal mines under the same rigorous inspections and regulations that covered coal mining. The people in the video agree that mining is much safer now than it was in 1972. They also agree that one of the reasons the fire was such a disaster was that no one believed it could happen. A hard-rock mine simply did not burn, so when it *did*, the miners initially refused to believe there was a problem. This video provides a stark look at a disaster and its aftermath by including stories about what happened, what went wrong, and what lessons were learned as a result. It also clearly illustrates how powerful a force culture can be when an event like this occurs.

During the taping of the Sunshine video, we captured many other stories. Miners shared stories about oldtimers who had trained them, about what it was like "back in the old days," and about close calls they had experienced or witnessed. One of the most poignant stories was told by Don Caparelli, who told us about the day Jimmy, his best friend and long-time partner,

died. Captured in the video *Hidden Scars*, Don tells about the day he and Jimmy were working in a raise that exploded in a violent rockburst. Don and Jimmy were thrown against the rib and buried with broken rock. Don had a couple of inches clear around his nose, but Jimmy was buried below him and could not breathe. Don talks about what it was like not to be able to move and to feel Jimmy struggle beneath him and finally die. Don waited for over 2 hours for rescuers to reach him, all the while believing that he, too, would suffocate as the rock tightened slowly around his chest with every breath he exhaled. Don says he has no physical scars to show for his ordeal, but the psychological ones, the hidden scars, he says will never heal.

Hidden Scars is about a mining fatality, but in a larger sense, it is about a workplace fatality. Don and Jimmy were friends, neighbors, and hunting buddies. There is a hole in Don's life that will never be filled, as there are holes in any organization that loses a cherished employee. This video, like *You Are My Sunshine*, is very relevant to nonmining industries, particularly those that involve danger and risk. In fact, both of these videos have migrated outside of mining and are being used in other industries that involve dangerous environments and constant risk.

Living with Loss: Surviving a Workplace Trauma

Five of the videos produced under this project, as well as the video on rock falls, were included in an independent evaluation study that sought to measure their effectiveness. There is an ongoing debate about how it is possible to isolate the effects of a single safety training video on overall safety in a mine, but the evaluators chose to ask the customers—trainers and trainees—themselves how they would rate the videos. Pre- and post-tests were created to gather quantitative and qualitative data on the videos with regard to both content and perceptions of effectiveness. A systematic study was done to collect data from a representative sample of trainers who used the videos and from miners both before they watched the videos and afterward.

The evaluation study (Fein & Isaacson, 2001) revealed five common themes. These themes did not appear to depend on whether the responses came from trainers or miners. These were—

Quality–The story lines were good, the information was important and well presented, and the technical quality of the videos, including sound and camera work, was excellent.

Credibility–Miners and trainers alike appreciated the use of real mines, real miners, and real situations to tell the stories. They felt a sense of respect for mining expressed in the videos.

Content–The information presented was specific rather than generic and tied to problems the miners faced every day. They particularly appreciated that the videos demonstrated both the wrong way to do things (which the new hires generally did quite well) and the right way, as shown by the experts.

Effectiveness–Trainers and miners alike believed the videos provided both good information and good reminders to workers, independent of whether they were new employees or experienced miners.

Engaging–The videos were easy to understand, made good use of humor or storytelling, and interesting. "No one sleeps through these" (Fein & Isaacson, 2001).

When we began the project, we identified six preliminary goals for the videos.

- Get the miners to pay attention in their training classes.
- Convince them that the information was important and relevant.
- Make the learning memorable.
- Make the learning enjoyable.
- Motivate the miners to choose to learn.
- Tie all training lessons to real-life situations.

From the information gathered during evaluation of the videos, it is apparent that the videos meet, at least to some extent, the goals of creating training materials that would capture miners' attention that miners would remember and that would in some way cause miners to choose to change their safety behavior or awareness. Miners and trainers do find the videos interesting; therefore, they not only pay attention, but according to the trainers, they actually ask if any new NIOSH videos are available. While the two videos that recount fatal accidents can not be described as "enjoyable" to watch, they have certainly proven to be memorable. Most important, perhaps, are the stories themselves. By watching peers tell stories about events that have happened because of inattention or errors in judgment, miners are drawn into the stories and have to reflect on what they would have done or how they would have felt. Responses show that these stories are very real to miners and have a significant impact on the decisions they make about safety (Fein & Isaacson, 2001).

Developing *effective* training can be an elusive goal. It is arguable by some that training is only effective if it changes someone's behavior or beliefs. Culture plays a key role in how people form their beliefs and subsequently in how they behave in relation to that belief system. What we learned in this project is that one must work inside the culture if one expects to affect behavior and that any permanent change in how miners view safety or their own role in working safely must come from within. Using stories, told by miners who look and talk like they do to teach lessons and reach their hearts, does work. In fact, it may be the only thing that works.

People make decisions with their heads, but they make life choices with their hearts. If the stories told by their peers convince them to work safely because it is the right thing to do, then they will voluntarily change their behavior. *That* is effective training.

REFERENCES

Bandura, A. A. (1997). *Self-efficacy: The exercise of control*. New York: Freeman.

Cullen, E. T. (2004). You are my Sunshine: The Sunshine Mine fire of May 2, 1972. Unpublished doctorial dissertation, Gonzaga University, Spokane, WA.

Fein, A.H. (2003). Evaluation of *Zen and the Art of Rock Bolting*. Report prepared for Spokane Research Laboratory, NIOSH, Spokane, WA. 38 pp.

Fein, A. H., & Isaacson, N. S. (2001). Video-based training program for underground miners evaluation report. Report prepared for Spokane Research Laboratory, NIOSH, 105 pp.

Gargiulo, T. L. (2002). *Making stories: A practical guide to organizational leaders and human response specialists*. Westport, CT: Quorum Books.

Knowles, M. S., Holton, E. F., & Swanson, R. A. (1998). *The adult learner* (5th ed.). Woburn, MA: Butterworth-Heinemann.

Livo, N. J., & Rietz, S. A. (1986). *Storytelling: Process and practice*. Littleton, CO: Libraries Unlimited, Inc.

Miller, H. B. (1991). *A comprehensive analysis of contract mining and wage incentives in underground metal mining*. Doctoral dissertation, Colorado School of Mines, Golden, CO.

Patton, M. Q. (2002). *Qualitative research and evaluation methods*. Thousand Oaks, CA: Sage.

Simmons, A. (2001). *The story factor: Inspiration, influence, and persuasion through the art of storytelling*. Cambridge, MA: Perseus Publishing.

Van Maanen, J., & Barley, S. R. (1984). Occupational communities: Culture and control in organizations. In: Staw & Cummings (Eds.), *Research in organizational behavior* (Vol. 6, pp. 287-366). Greenwich, CT: JAI Press.

Wlodowski, R. J. (1985). *Enhancing adult motivation to learn*. San Francisco: Jossey-Bass.

Zemke, R. (2002). Who needs learning theory anyway? *Training*, 39(9), 86-91.

UNDERSTANDING SELF IN STRESSFUL WORKING ENVIRONMENTS

Thomas W. Camm[6]

There are also substantial noneconomic consequences of workplace injuries and illnesses on quality of life. Physical and psychological functioning in everyday activities may be affected, self-esteem and self-conf dence may be reduced, and an individual' s role in the family and community may change. Even less research has been focused on these nonmonetary costs. Studies of unemployed workers and their families and of people with chronic illnesses and disabling injuries show that income and employment losses, illness, and physical impairment can have profound human consequences on both workers and their families. Better measures of both economic impacts (direct and indirect) and noneconomic impacts will help improve targeting of resources for research, prevention, and compensation. (NIOSH, 2002)

INTRODUCTION

Millions of occupational injuries and thousands of occupational fatalities occur each year in the United States. While fatalities are the most dramatic and tragic, nonfatal injuries may still have a devastating impact on families. However, an injury and its attendant medical and labor market consequences are an incomplete description of the economic consequences for the worker and his or her family. In particular, the changed economic circumstances of the family and possible increased care required for the injured worker may affect economic and social outcomes and behaviors for other family members, including children.

Thus direct medical costs of an injury, lost wages, and health care are often a small percentage of the total impacts. Numerous indirect and unrecognized costs are associated with workplace injuries: reduced income, depletion of savings, and loss of assets (which could include automobiles or even homes). Potential costs to workers that may not be obvious are the need for professional counseling, caregiver services in the home, home modifications and equipment related to disability, and deferral or loss of education opportunities for family members.

Some of the major direct costs of a workplace injury for an employer include medical expenses, legal and administrative expenses, worker's compensation administration, and property damage. Indirect costs include the need for retraining and/or re-staffing, disruption of work processes, and the effects of workplace injury, exposure, or fatality on the productivity of co-workers who see themselves as being at a heightened risk of injury (Camm, 2000).

These consequences are all the focus of a growing body of literature (Leigh et al., 1997; Miller, 1997; Miller & Galbraith, 1995; Viscusi, 1996). Extending the literature on earnings losses for workers with injuries to losses in family income is an area of increasing interest to occupational economics researchers.

Nonetheless, little attention has been given to the social costs of an injury. The ability of workers to function in an occupational setting is contingent on their individual emotional health, acute or chronic forms of depression, anxiety, or inappropriate anger, and can have a significant impact on each worker's ability to perform tasks.

---

[6]Mining engineer, Spokane Research Laboratory, NIOSH, Spokane, WA.

Forty percent of workers reported their jobs as "very" or "extremely stressful" in a survey by Northwestern National Life (National Institute for Occupational Safety and Health [NIOSH], 1999).

Heightened levels of stress associated with high-risk occupations or working conditions, particularly in a setting where a recent serious injury or a fatality occurred, can have significant impacts to both the productivity and long-term health of workers. Symptoms of stress can be physical, mental, and/or behavioral. Mental and behavioral effects include depression, anger, and anxiety (Freudenberger, 1998). The effects of low cognitive levels as a result of high stress levels and unsafe/unhealthy working conditions (whether real or perceived) can also adversely affect a worker's ability to pay attention to work, be aware of at-risk behavior, and work productively.

The fourth edition of the Diagnostic and Statistical Manual of Mental Disorders (DSM-IV) (American Psychological Association [APA], 1994) notes that research shows a symptomatic link between increased levels of depression, anxiety, and anger, and a lack of occupational and social functioning. The DSM-IV identifies symptoms of such emotional dysfunction extreme enough to cause "clinically significant distress or impairment in social, occupational, or other important areas of functioning" (pp. 327, 436, 646).

NIOSH researchers (2002) noted that health care expenditures were nearly 50% greater for workers who reported high levels of stress. Current research in occupational health psychology indicates that high levels of stress lead to employees feeling less in control of their workplace, resulting in negative health effects on individual workers (Sauter et al., 1999).

SYSTEMS THINKING

Systems and *systems thinking* are terms used frequently in many disciplines. However, an ecologist using the term *ecosystem* or *ecological system* is describing a system in a different way than a physicist or an engineer. A psychologist talking about a family system will be discussing ideas that in some ways are similar, but in others quite different, from an organizational theorist discussing the system of a business organization. The following sections discuss general systems theory as the overarching theory for systems and then focus on specialized aspects of general systems theory: systems engineering and Bowen's family systems theory. A systems approach can provide a useful framework for an integrated view of the consequences of stress in the workplace.

General systems theory

The root meaning of the word "system" is the Greek *synhistanai*, which literally means "to place together." Understanding things systemically means putting them into context to establish the nature of their relationships.

One definition of a *system* is "an integrated whole whose essential properties arise from the relationships between its parts, and 'systems thinking' the understanding of a phenomenon within the context of a larger whole" (Capra, 1996, p. 27).

A key characteristic of the organization of living organisms is the tendency to form multi-leveled structures of systems within systems, referred to as *hierarchies*. Hierarchy, in this sense, has a different meaning than is typically thought of in organizations; in nature, there is no "above and below" so much as networks within other networks. Connectedness, relationships, and context are fundamental to understanding systems theory. In systems theory, the interactions and relationships among parts of a system are as important as their individual characteristics in understanding the dynamics of the sys-tem.

Von Bertalanffy was one of the first to establish systems thinking as a major scientific movement (Capra, 1996). According to von Bertalanffy (1968), general systems theory is a general science of "wholeness" with an emphasis on the use of mathematics to define principles that apply to systems in general. "It attempts to define principles found universally in all systems in nature" (Papero, 1990, p. 3).

Von Bertalanffy distinguishes between "closed systems" (systems considered to be isolated from their environment) and "open systems." Physical chemistry and thermodynamics are examples of closed systems. "However, we find [that] systems by their very nature and definition are not closed systems. Every living organism is essentially an open system," maintaining itself in a continuous inflow and outflow, building up and breaking down of components, never in a state of chemical or thermodynamic equilibrium (von Bertalanffy, 1968, p. 39).

Although a living system is never in a state of perfect equilibrium, it is constantly seeking to reach such a state. A system's stability is continually tested by fluctuations, which can cause a deviation that can be either corrected or magnified by positive or negative feedback. "Homeostasis" is the systems term for describing this state of dynamic balance, which is characterized by multiple, interdependent fluctuations (Capra, 1982, pp. 286-287).

David Bohm used the term "rheomode" (rheo from Greek "to flow") to describe the interconnected aspect of systems theory. Bohm was a quantum physicist who saw that Western civilization's tendency to describe the universe as discreet fragments provided an inadequate world view. Rheomode is a way to describe the universe so that, in essence, everything in the universe is unbroken and undivided whole movement. To attempt to describe the universe as fragments misses the essence of the connected nature of the universe. "Rather, it [rheomode] implies that any describable event, object, entity, etc., is an abstraction from an unknown and undefinable totality of flowing movement" (Bohm, 1980, p. 49). Bohm used the image of a flowing stream, an ever-changing pattern of waves and ripples that have no independent existence apart from

the stream. Applying this principle to human health, Bohm stressed that the fragmented nature of current culture, with its emphasis on autonomy, has adverse effects.

The paradox of human nature is that society has a sense that wholeness or integrity is necessary to make life worth living, yet most social structures emphasize a fragmented existence. The word "health" in English is based on the Anglo-Saxon "hale," meaning whole—to be healthy is to be whole (Bohm, 1980). The totality of existence is an unbroken wholeness, an undivided flowing movement without borders.

Systems engineering

Systems engineering is an interdisciplinary approach that helps enable the realization of successful systems. It focuses on defining customer needs and required functionality early in the development cycle, documents requirements, then proceeds with design synthesis and system validation while considering the complete problem, which may include operations, performance, testing, manufacturing, costs and scheduling, training and support, and disposal.

Systems engineering integrates all disciplines and specialty groups into a team effort to form a structured development process that proceeds from concept to production to operation. Both business and technical needs are considered, all with a goal of providing a quality product that meets customer needs (International Council on Systems Engineering [INCOSE], 1999).

A safety system model can be used that includes four main components: economics, engineering, work environment, and human factors (figure 1). These components provide a model based on viewing the safety of a work organization as an overall system (Camm & Girard-Dwyer, 2004). Using principles from systems theory and systems engineering, this approach allows an injury or fatality to be evaluated from multiple perspectives.

1. Economics considers direct, indirect, productivity, and intangible costs.
2. Engineering considers the design and types of equipment, ergonomics, technological complexity, work process, cost effectiveness, and maintenance.
3. Work environment includes physical agents, chemicals, noise, dust, and visibility.
4. Human factors includes physical capability perceptual motor skills and abilities, intellectual aptitude, personality, training, and work design.

Human factors in systems engineering emphasizes that the performance of a machine or system depends on the operator, not just the equipment. Individual workers in a poorly designed system will often continue to function, but this usually results in increases in training time, stress for operators, and human errors, plus under-use of equipment capabilities (Chapanis, 1996, p. 19). Using this four-part systems model allows researchers to focus on key areas and interrelationships of the overall safety system of a work setting.

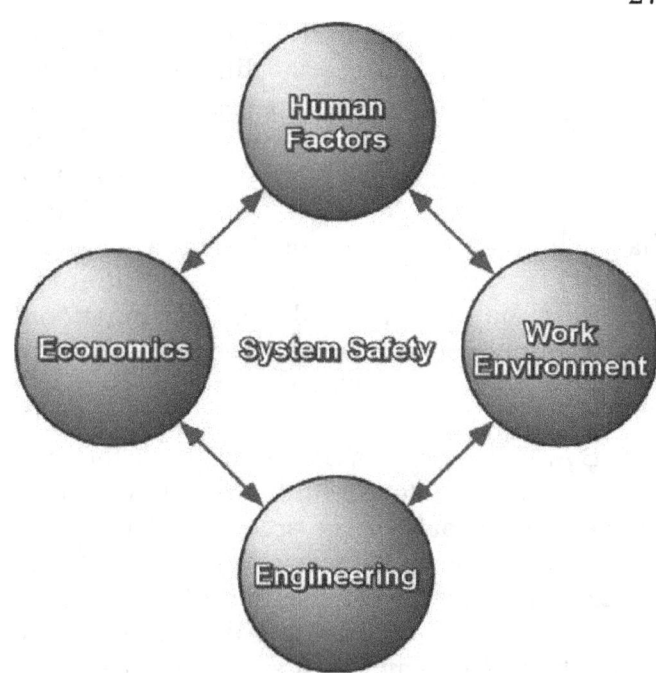

Figure 1.—Model of factors used in systems engineering

BOWEN'S FAMILY SYSTEMS THEORY

Rigid hierarchies are often the bane of followers in an organization, depriving them of the opportunity to make meaningful contributions to a system and locking them into an ongoing subservient role. A typical organizational system tends to maintain a hierarchy or chain of command by utilizing varying degrees of rigidity. According to systems theories (Bowen, 1978; Satir, 1964), as the social interactions within a system become more rigid, the system itself becomes less functional.

Bowen (1978) framed the quality of communication (or lack of communication) in the hierarchy of a system as indicative of its developmental level or maturity. Papero (1995) pointed to Bowen's conceptualization of three equally important therapeutic goals through focusing on communication patterns in a family system:

1. Building person-to-person relationships,
2. Becoming a better observer of self and others, and
3. Detriangling," or disengaging oneself from emotionally laden or "stuck" situations.

From Bowen's perspective,

[A] person-to-person relationship is one in which two people can relate personally to each other about each other , without talking about others (triangling), and without talking about impersonal "things." Few people can talk personally to anyone for more than a few minutes without increasing anxiety , which results in silences, talking about others, or talking about impersonal things. (1978, p. 540)

Bowen's Family Systems theory provides four interlocking concepts to form the foundation of understanding effectiveness in the workplace: differentiation of self, triangles, emotional cutoff, and societal regression. Each of these concepts focuses attention on the central concept of workplace interaction.

Differentiation of self

Differentiation of self is the cornerstone of Bowen's theory. Differentiation can be a significant component of an individual's effectiveness in a work setting because the differentiated person can relate consistently on a person-to-person level without digressing toward emotional cutoff, fusion, triangling, or mundane interaction. An undifferentiated person, on the other hand, has achieved little emotional separation from others. The greater the degree of undifferentiated self, the greater the likelihood of emotional fusion into a common self with others (Goldenberg & Goldenberg, 2000).

Bowen developed a theoretical scale for evaluating a person's level of differentiation; values range from 0 to 100. The scale is not designed to assign people to an exact level, but rather to describe relative levels of differentiation. Complete undifferentiation exists in a person who has achieved no emotional separation from the system (family or work organization) in which he or she resides. "He is a 'no-self,' incapable of being an individual in a group. This level of functioning is arbitrarily assigned a scale value of 0" (Kerr & Bowen, 1988, p. 97).

Someone who has achieved complete differentiation has fully resolved the inappropriate (too close or too distant) emotional attachment to his or her system (family, and/or, by extension, work relationships) and would be arbitrarily assigned a scale value of 100. Complete differentiation is characterized by someone who has attained complete emotional maturity that allows him or her to be an individual in a group and attain personal responsibility without fostering or participating in the irresponsibility of others.

Bowen used this differentiation scale as a theoretical standard, stating that no one could actually attain the level of 100. The scale is not intended to be either a diagnosis or an instrument for assigning exact levels to individuals; rather, it is intended to define an individual's ability to adapt to stress. The higher the level of differentiation, the more stress required to trigger an unhealthy symptom (Kerr & Bowen, 1988).

According to Bowen (1978), the key characteristic describing the difference among people is the degree to which individuals can distinguish between the feeling process and the intellectual process. What distinguishes someone at a higher level of differentiation is the ability to choose to have his or her functioning guided by connection-oriented thoughts rather than overly "fused" or overly distant feelings. Individuals with low differentiation have their emotions and intellect so enmeshed that their lives are dominated by the feelings of those around them. In the work setting, this low differentiation can lead to reluctance on the part of workers to speak out when they see unsafe working conditions or practices and put themselves at risk rather than face the possibility of real or perceived sanctions by co-workers or managers.

When applied to the workplace, Bowen's theory provides a framework for describing a safer, more meaningful work environment. If a worker feels pressure to conform to the demands of a group or a boss, the worker may compromise his or her own safety to avoid appearing weak or unproductive. This dynamic may happen even when no expectation to perform a work function in an unsafe manner exists and when this perception is held by the worker only.

Systems theory provides a framework to describe the human capacity for cohesiveness, altruism, and cooperativeness—all characteristics that contribute to a safe, efficient work setting. "The higher the level of differentiation of people in a family or other social group, the more they can cooperate, look out for one another's welfare, and stay in adequate contact during stressful as well as calm periods" (Kerr & Bowen, 1988, p. 93). Bowen's theory indicates that a worker with a higher differentiation of self is more likely to respond in a thoughtful manner to an unsafe situation, to handle the stress of an injury or fatality at the workplace, or to respond to any of various stress-producing experiences in a more healthy, constructive way.

Triangles

The triangle is the basic building block in an emotional system, according to Bowen (Kerr & Bowen, 1988). A two-person system is inherently unstable. Each person brings in a third under stressful conditions as each attempts to create a triangle to reduce increasing stress in the two-person system (Goldenberg & Goldenberg, 2000). During periods of low intensity, two-person systems can function at a comfortable level, but when the stability of the situation is threatened by one or both persons (which increases anxiety levels), at least one of them will involve a vulnerable third party. By reaching out and pulling a third person into a triangle relationship, the anxiety in the system is diluted. The triangle is both more stable and flexible than a twosome and has a higher tolerance for dealing with stress (Bowen, 1978).

Anxiety is the major influence on the activity of a triangle. An increase in anxiety will erode the level of comfort for individuals in a triangle. This discomfort is typically felt more by one person in the triangle than others (Kerr & Bowen, 1988). During high-stress periods, the emotional processes in the triangle can undergo changes. For instance, person A and person B can be a twosome that functions with a certain level of comfort. When a stressful situation arises, person B is uncomfortable with how A responds and forms an emotional bond of support with person C, which creates a triangle. Because B is now getting emotional support from C, neither B nor C need A for the same emotional bond, which could cause a widening gap in the bond between A and B. C helps B to maintain a distance from A that was not possible in the original twosome. This changes the character of the original twosome and can lead to the result that B (who was previously uncomfortable) is now comfortable, although the increased distance may cause A to go from a position of comfort to discomfort, all as a result of a triangle being formed.

Maintaining an emotionally neutral attitude plays an important part in avoiding the formation of harmful triangles. "The more it is possible to be emotionally neutral about the rela-tionship process between others, the more effective will be any detriangling maneuver" (Kerr & Bowen, 1988, p. 150). Detriangling efforts often do not go smoothly; consequently, the more intense the emotional situation, the more important it is for a person's thinking (and corresponding behavior) to be tactful, under reasonable control, and focused on long-term relational quality. If an uptight person who is not neutral attempts to detriangle from a highly anxious situation or system, it is likely that the person will inadvertently make the problem worse.

Emotional cutoff

"The most obvious manifestation of the emotional system is the reactivity of the individual to its environment" (Papero, 1990, p. 45). The higher the level of chronic anxiety in a setting, the more each individual will tend to react emotionally to reduce the anxiety (p. 43). When individuals have difficulty dealing with intense relationships or social situations (including a work setting), an emotional cutoff allows a way to deal with the situation by distancing themselves from the relationships involved and maintaining a relatively fixed distance from others (pp. 62-63).

Forming triangles and emotional cutoff can be related. As an example, two miners might be having a conflict that leads to argument and animosity when a shift boss approaches them and tells them to get back to work. To relieve the anxiety between the two of them, when the boss leaves the two workers turn their animosity toward talking disparagingly about the boss, making themselves allies and relieving the ten-sion between themselves. This is a typical scenario of triangling a third party in to relieve tension between two people. Left unresolved, this tension with the boss could lead a worker to feel disconnected and eventually to distance him- or herself from the boss in particular and management in general. Emotional cutoff by workers can then make it extremely difficult to train workers, including training them to work in a safer manner.

Societal regression

Bowen (1978) extended his theory to include the impact of individual, familial, or work systems on societal systems as a whole. Societal regression represents Bowen's thinking on society's emotional, social, and occupational functioning. He argued that society (as individuals and work systems) is caught between the opposing forces of undifferentiation and differentiation. When chronic stress (e.g., population growth, limited economic resources, depletion of natural resources) is exacerbated by continual acute stressors (dysfunctional individuals, dysfunctional systems), the result is likely to be a surge of destructive patterns (such as increased triangulation and emotional cutoff) that then result in the continual erosion of society as reflected in chronically sick organizations. In the end, this bleak progression leads to deepening emotional dysfunction and eventual emotional breakdown among individuals and is characterized by more tenacious forms of depression, anxiety, and anger.

"Heightened anxiety in society produces a surge of togetherness, which in turn creates greater discomfort and further anxiety" (Papero, 1990, p. 64). It is not possible to reduce anxiety in a social setting without reducing the emotionality that propels it and the consequent problems of cutoff and enmeshed triangles. Only the presence of well-differentiated individuals in the system can alleviate the symptoms of reactivity to this environment of heightened anxiety.

JOB STRESS

"Job stress can be defined as the harmful physical and emotional responses that occur when the requirements of the job do not match the capabilities, resources, or needs of the worker. Job stress can lead to poor health and even injury" (NIOSH, 1999, p. 6). In addition to the direct costs to the company and the worker, the social costs and consequences of job stress can include depression, anger, anxiety, public assistance, reduced family income, strained relationships (conflict, divorce, etc.), and a reduction in community involvement. "The American Institute of Stress…estimates that stress and the ills it can cause—absenteeism, burnout, mental health problems—costs American business more than $300 billion a year" (Daniels, 2002).

Signs of job stress include headache, sleep disturbances, difficulty in concentrating, short temper, upset stomach, job dissatisfaction, and low morale. Chronic job stress can increase the risk of health problems—cardiovascular disease, musculoskeletal disorders, psychological disorders, workplace injury, suicide, ulcers, and im-paired immune function.

Many approaches are available to reducing job stress: making sure the workload matches capabilities and resources, designing jobs to provide meaning and opportunities, clearly defining roles and responsibilities, providing opportunities for workers to participate in decisions affecting their jobs, improving communications, reducing uncertainty, enhancing opportunities for social interaction among workers, and creating work schedules that are compatible with demands and responsibilities outside the job (NIOSH, 1999).

People—and systems—have limits beyond which they experience a serious breakdown in performance. "Human beings are systems, and as such have physical, performance, emotional, and mental limits" (Swenson, 1992, p. 74). Swenson describes a simple equation to evaluate the level of stress a person is experiencing, using the term "margin" to describe the gap between rest and exhaustion, the opposite of overload.

$$Margin = Power - Load$$

Power is made up of factors that include skills, time, emotional stability, physical strength, social supports, financial security, spiritual vitality, education, and training. Load combines internal factors like personal expectations and emotional disabilities; and external factors like work, relationship issues and responsibilities, financial debts, and health problems.

When our load is greater than our power we enter into negative margin status, that is, we are overloaded. Endured long-term, this is not a healthy state. Severe negative margin for an extended period of time is another name for burnout. When our power is greater than the load, however , we have margin. (Swenson, 1992, p. 92)

As shown in the human function curve in figure 2 (Swenson, 1992), a moderate increase in stress can lead to an increase in productivity. However, every individual has a point at which they reach maximum productivity (point B), Beyond this point, any increase in stress leads to a decrease in productivity. The dilemma for an organization is knowing the difference between when a workforce is at point A (with the potential for an increase in productivity) and when it is at point C (burnout). Since both points A and C have similar productivity levels, it is often difficult to determine where on the curve a workforce (or an individual) may be at any given time.

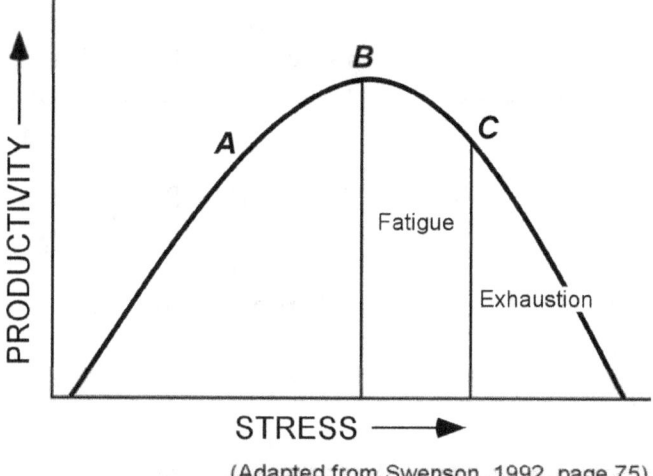

(Adapted from Swenson, 1992, page 75)

Figure 2.—Human function curve

Understanding the dynamics of how individ-uals react to stress in a work environment will help to develop appropriate work processes and training to keep stress at acceptable levels or, to use Swenson's terminology, to keep the workplace in positive margin and avoid worker burnout. The concept of differentiation provides a theoretical framework for understanding the individual, how people differ in terms of sensitivity to one another, and their ability to preserve autonomy in the face of pressures for togetherness (Papero, 1990, p. 45).

Understanding the reactivity of the individual worker to the work environment and the ability to keep thinking and emotional systems separate can lead to a clear thought process that incorporates complexity and avoids simplistic efforts to fix blame or causality. "The higher the level of differentiation of people in a family or other social group, the more they can cooperate, look out for one another's welfare, and stay in adequate contact during stressful as well as calm periods" (Kerr & Bowen, 1988, p. 93).

Triangles become more active where individuals have a greater degree of undifferentiation and the system has intense anxiety. Two-person relationships are essentially unstable when tense or anxious, with anxiety often being more intense in one person than the other. When anxiety exceeds a tolerable level, a third person is involved, forming a triangle. When anxiety is high throughout the system, the tension spreads through webs of interlocking triangles (Papero, 1990, p. 49-51). "The major influence on the activity of a triangle is anxiety" (Kerr & Bowen, 1988, p. 135).

Every individual has a threshold beyond which additional stress makes the situation catastrophic. "The capacity to bear anxiety is important for the individual's self-realization and for his conquest of his environment" (May, 1977, p. 65). When an individual has a weak sense of self and is overwhelmed by the adverse affects of triangle, one response is to reduce or totally cut off emotional contact. "Emotional cutoff is at a minimum when people consistently act toward one another on the basis of mutual respect and are able to listen to one another without emotional reactivity interfering with the ability to 'hear' each other's thoughts and feelings" (Kerr & Bowen, 1988, p. 325).

When the dynamics of understanding the work setting as a system are understood, the characteristics of human interaction described by systems theory provide an approach to minimize the amount of unhealthy chronic stress in a work environment.

Research on application of Bowen Family Systems in the workplace has been mostly qualitative (Comella, et al., 1996; Sagar & Wiseman, 1982). Further empirical research would provide additional data for applications in work organizations.

It should be noted that when employees are experiencing emotional distress that affects their work performance, professional help from an employee assistance program (EAP) counseling service is often appropriate. If the primary source of a worker's emotional distress is related to the structure and functioning of the work organization, then it is appropriate for management to take steps to make changes.

REFERENCES

American Psychiatric Association. (1994). *Diagnostic and statistical manual of mental disorders* (4th ed.). Washington, D.C.

von Bertalanffy, L. (1968). *General system theory: Foundations, development, applications* (Rev. ed.). New York: George Braziller.

Bohm, D. (1980). *Wholeness and the implicate order*. London: Routledge.

Bowen, M. (1978). *Family therapy in clinical practice*. Northvale, NJ: Jason Aronson.

Camm, T. W. (2000). *Economics of safety at surface mine spoil piles*. Washington, DC: Department of Health and Human Services, National Institute for Occupational Safety and Health, NIOSH RI 9653, DHHS (NIOSH) Publication No. 2000-129.

Camm, T. & Girard-Dwyer, J. (2004). *Economic consequences of mining injuries*. SME Preprint no. 04-37.

Capra, F. (1982). *The turning point: Science, society, and the rising culture*. Toronto: Bantam Books.

Capra, F. (1996). *The web of life*. New York: Anchor Books.

Chapanis, A. (1996). *Human factors in systems engineering*. New York: John Wiley & Sons, Inc.

Daniels, C. (2002, October 28). The last taboo: It's not sex. It's not drinking. It's stress—and it's soaring. *Fortune.com*. Retrieved October 21, 2002, from http://www.fortune.com/indext. jhtml?channel=print_article.jhtml&doc_id=209802

Comella, P. A., Bader, J., Ball, J. S., Wiseman, K. K. & Sagar, R. R. (1996). *The emotional side of organizations: Applications of Bowen Theory*. Washington, DC: Georgetown Family Center.

Freudenberger, H. J. (1998). Stress and burnout and their implication in the work environment. In J. M. Stellman (Ed.), *Encyclopaedia of occupational health and safety* (4th ed., Vol. 1, pp. 5.15-5.17). Geneva, Switzerland: International Labour Office.

Goldenberg, I. & Goldenberg, H. (2000). *Family therapy: An overview* (5th ed.). Belmont, CA: Brooks/Cole.

INCOSE [International Council on Systems Engineering]. (1999, June 4). *What is systems engineering?* Retrieved May 6, 2002, from: http://www.incose.org/whatis.html.

Kerr, M. E. & Bowen, M. (1988). *Family evaluation: An approach based on Bowen theory*. New York: W. W. Norton & Co.

Leigh, J. P., Markowitz, S. B., Fahs, M., Shin, C., & Landrigan, P. J. (1997, July 28). Occupational injury and illness in the United States. *Archives of Internal Medicine*, 157, 1557-1568.

May, R. (1977). *The meaning of anxiety*. New York: W. W. Norton & Co.

Miller, T. R. (1997). Estimating the costs of injury to U.S. employers. *J. Saf. Res.*, 28,no. 1, 1-13.

Miller, T. R. & Galbraith, M. (1995). Estimating the costs of occupational injury in the United States. *Accident analysis and prevention*, 27, no. 6, 741-747.

NIOSH. (2002). *NORA Social and economic consequences additional information*. Retrieved May 9, 2002, from http://www2.cdc.gov/NORA/NaddinfoSocio.html.

NIOSH. (1999). *Stress at work*. Washington, DC: Department of Health and Human Services, National Institute for Occupational Safety and Health, DHHS (NIOSH) Publication No. 99-101.

Papero, D. V. (1990). *Bowen family systems theory*. Boston: Allyn and Bacon.

Papero, D. V. (1995). Bowen family systems and marriage. In N. S. Jacobson & A. S. Gurman (Eds.), *Clinical handbook of couple therapy*. New York, NY: The Guilford Press.

Sagar, R. R. & Wiseman, K. K. (1982). *Understanding organizations: Applications of Bowen Family Systems theory*. Washington, DC: Georgetown Family Center.

Satir, V. (1964). *Conjoint family therapy*. Palo Alto, CA: Sciences and Behavioral Books.

Sauter, S. L., Hurrell, J. J., Jr., Fox, H. R., Terrick, L. E., & Barling, J. (1999). Occupational health psychology: An emerging discipline. *Industrial Health*, 37, 199-211.

Swenson, R. A. (1992). *Margin*. Colorado Springs, CO: NavPress.

Viscusi, W. K. (1996). The dangers of unbounded commitments to regulate risk. In: R. W. Hahn (Ed.), *Risks, costs, and lives saved: Getting better results from regulation*. New York: Oxford Univ. Press., pp. 135-166.